All About Sugar

Do you have a sweet tooth?

THE WORLD OF FOOD

Fruit becomes sweeter and more sugary as it ripens in the sun.

You know what sugar looks like and tastes like. If you have a sweet tooth, you probably like it on your cereal, or fruit pie. But did you know that there is sugar in lots of plants, even though we can't see it?

Where is the sugar and why can't we see it?

Every plant that has leaves contains sugar. It is stored in the sap of trees, in fruit and in the roots, leaves and stalks of plants. Some plants contain much more sugar than others. Try a bite of banana, followed by a carrot stick. Which is sweeter?

How does the sugar get there?

Chlorophyll, which is what makes green leaves green, absorbs the sun's rays and combines with water and air to make sugar.

Bees collect a sugary liquid called nectar from flowers and change it into honey, which they store in the hive.

It was a long time before people discovered how to get the sugar out of plants. What did they use instead? **Honey from bees.** In those days, honey was so precious that half a litre cost as much as a whole sheep!

Bees' picnic

In Egypt, beekeepers still load their beehives on to boats and sail slowly down the Nile with them, as they did in ancient times. The bees buzz to and fro, collecting nectar from the flowers that grow all along the river banks. Because there are so many more flowers for them to feed on, they produce much more honey than they would have done if they had stayed at home.

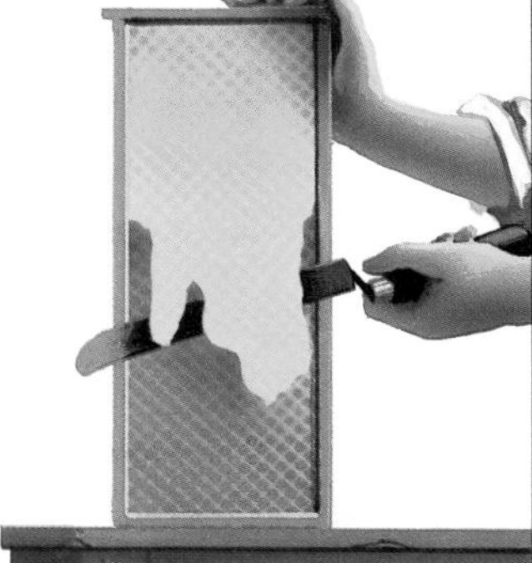

To collect the honey, the beekeeper takes the combs out of the hive, he scrapes them and the honey oozes out.

Where was sugar first discovered?
In India, a special sort of reed used to grow wild. The Indians found a way of squeezing the juice out of the reed and turning it into sugar crystals. The word sugar comes from the Sanskrit word 'sarkara', meaning sugar crystals.
The reed they found growing wild was sugar cane. When the Persians invaded India in the 6th century B.C., they found sugar cane growing there and they soon became experts in growing and making sugar. By the 11th century, merchants had brought sugar to Europe.
In those days, sugar, like pepper and cinnamon, was still rare. People had to use it sparingly because it was so expensive. One use was in medicines to make them taste less bitter.

When Alexander the Great conquered Asia in the 4th century B.C., the Greeks and Romans began importing sugar.

Where does sugar cane grow?

It grows in tropical countries, where the climate is hot and humid. Inside the stem of the sugar cane there is a firm, white pulp, which is where the sugar is stored.

Planting sugar cane

Short sections of sugar cane, called 'setts', are laid flat on the ground in

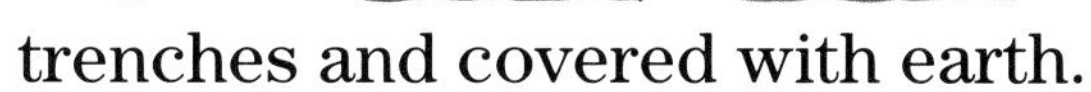

trenches and covered with earth. These setts take root and send up shoots, which grow into canes.

Eighteen months later, the canes are ready to be harvested. The following year, fresh canes grow from the old ones, and are cut. This goes on for another five or six years. Then it is time to take up the old setts and plant new ones.

Strip off the hard outer skin and take a bite. The sugar cane pulp is deliciously sweet and chewy.

The soil must be ploughed before the setts are laid.

Harvesting the cane

When they are 4 or 5 metres high, the canes are ready to be cut. Nowadays there are machines which will do this job, but in many places the canes are still cut by hand, which is very hard work.

Sugar cane doesn't keep once it's cut!

As soon as the canes have been cut, they begin to lose their sweetness. So they are picked up as quickly as possible and taken to the sugar factory.

Laying the setts on the ploughed earth

The sugar may be poured into cone-shaped moulds while it is still hot.

The sugar mill

In the past canes were crushed in a mill. As the mill ground them down, a dark brown liquid, full of little bits of leaf and stem, came out.

How is the sugar extracted?
The dirty brown liquid is heated, and the dirt is removed. The liquid thickens, and **after a while, sugar crystals appear.** These are separated from the syrup by spinning it round and round at high speed. Traditional methods are now being mechanised in most parts of the world.

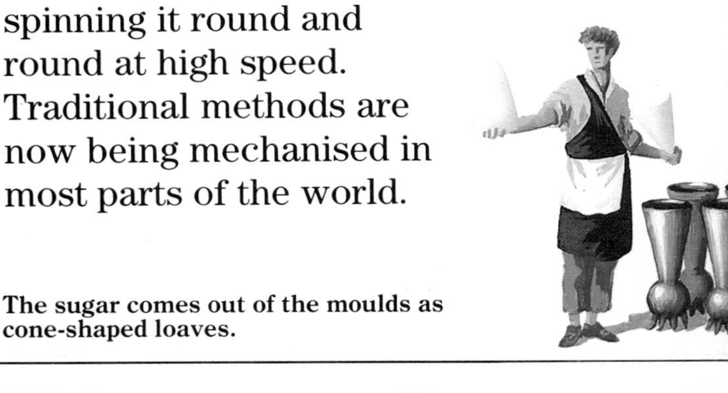

The sugar comes out of the moulds as cone-shaped loaves.

In Europe, chocolate and coffee were becoming more and more popular, and to go with them, more and more sugar was in demand.

In 1807, England was at war with France. All overseas trade came to a standstill. **The ships bringing the sugar cane couldn't get through!** The English had developed a sweet tooth and so had the French. How would they manage now there was no more sugar coming in?

Sugar beet is a white root vegetable, rather like a fat parsnip.

The answer was there all the time in the ground – sugar beet

This root vegetable grows well in mild climates like ours. As it grows, it stores sugar in its root below the ground. People experimented with carrots, grapes and even potatoes to see which would produce the most sugar, but sugar beet was found to be by far the best. Sugar was first extracted from sugar beet in 1747.

A French chemist found a way of getting the sugar out of sugar beet.

His name was Benjamin Delessert. Napoleon was so pleased that he ordered huge fields of sugar beet to be planted, and soon the French had all the sugar they wanted.

Napoleon decorated Delessert for his discovery.

Sugar beet: From seed to sugar lump

1. The sugar beet seed is planted in the Spring. By May, clumps of green leaves are showing above the ground.

3. The harvesting machine hauls the beets out of the ground, cleans them and drops them into a trailer.

5. The beets are washed and sliced, and put into hot water so that the sugar can be extracted.

2. In the Autumn, the beets are ready to harvest! First, the leaves are cut off by a special harvesting machine. They are used for animal feed.

4. The beet are then taken to the nearby beet factory. From September until Christmas these factories are at work day and night making sugar.

6. The sugar juice is heated and all the dirt is removed. Then the liquid is boiled until only sugar crystals are left.

4
3
2
3
1
2
3
3
3
1
3
3

There are all sorts of different sugars to choose from. Some are coarse, others are fine; some are white, others are brown. **White granulated sugar is more widely used than any other sort.**

Brown sugars can be any shade of brown, from light brown to very dark brown. They have a stronger taste than white sugar. Which do you prefer? Demerara, soft brown, dark brown or Muscovado?

All sugar crystals are the same shape. In some sugars, they are all small (1). Sugar candy is made of bigger sugar crystals which take longer to form (2). Sugar lumps are crystallised sugar which has been moulded into different shapes and sizes (3). Caster sugar (4) is crystallised sugar which has been ground up. Icing sugar is finely ground caster sugar.

In Arab countries, you can still buy sugar in loaves, or sections of loaves.

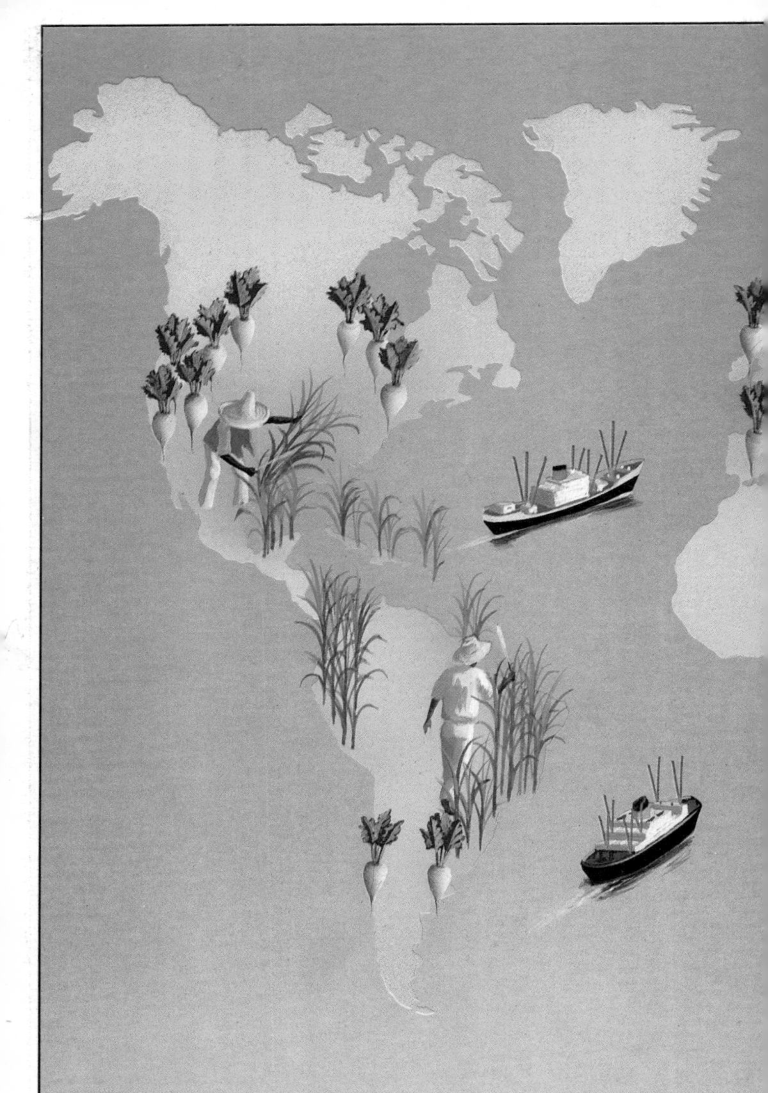

This map shows which parts of the world grow sugar cane and which grow sugar beet.

In Britain, half the sugar we eat comes from sugar beet which we grow here. The rest is made from sugar cane which we buy from other countries.

Most sugar comes from sugar cane and sugar beet, but some comes from other plants.

The Canadians make maple syrup from the sap of the maple tree. In March, when the sap begins to rise in the tree, it oozes out of small cuts made in the tree trunks and is collected in little buckets. One tree can produce as much as 10 litres of sap in a day. The sap is boiled until only the delicious runny syrup remains.

In the United States, more and more sugar is made from maize. In Pakistan, sugar is made from sweet, sticky dates.

A date palm

Three hundred years ago you could only buy 30 grammes of sugar at a time: that's about two heaped tablespoonfuls! Sugar was kept locked up in a box, and the father of the family carried the key around with him for safekeeping. Nowadays we buy sugar by the kilo. In Britain, we each eat an average 36-37 kilogrammes a year! That's probably more than you weigh!

Is sugar good for us?

If you are playing lots of sport, sugar can give you a quick boost of energy when you are tired. But, don't forget that it can make you fat if you eat too much of it, and that sugar and sweets can also make your teeth rot. So always clean your teeth and visit your dentist. And remember if you have a sweet tooth that too much sugar is a bad thing.

Sweet things like lollipops, ice cream, cakes, fizzy drinks and sweets are made from sugar.

Wasps feed on the juice of ripe blackberries.

You are not the only one with a sweet tooth!

You are just tucking into a picnic on a hot summer's day, when along comes a wasp or a bee and settles on your cake. How you wish it would go away!

All sorts of animals, big and small, like the sweet taste of fruit, and come from far and wide for a quick peck or nibble as it ripens. Rats are a nuisance in fields of sugar cane. They bite holes in the bottom of the canes and suck out the juice.

Bears love honey! If they spot a bees' nest, they knock it out of the tree and help themselves to the honey, gobbling it up in great pawfuls.

Blackbirds are very fond of ripe fruit. If you are not careful, they will help themselves to the strawberries in your garden. Most people put nets up to keep them away from the ripening fruit.

Have you ever made marzipan?

You need: 250 grammes of icing sugar, 250 grammes of ground almonds, 2 egg whites, some food colouring (red, green, yellow).

Mix the icing sugar and the ground almonds in a bowl. Add the egg whites and colouring and mix until you have a firm, stretchy paste that is easy to work. Leave the paste for an hour or two to dry out.

Marzipan animals taste as good as they look!

Collect together some prunes, dates, glacé cherries, almonds, walnuts, hazelnuts, sticks of liquorice, and away you go.

Marzipan mouse: Take a little marzipan and mould it into a mouse shape. Use two almonds for its ears, and liquorice for its eyes and tail.

Ladybird: The body is a hazelnut wrapped in marzipan; the spots and antennae are liquorice.

You'll be able to think of lots more animals like these to make.

Index

animals, 33
bears, 33
bees, 9, 33
birds, 33
brown sugar, 25
candy, 25
caster sugar, 25
chlorophyll, 7
chocolate, 19
coffee, 19
dates, 29
Delessert, Benjamin, 21
demerara, 25
harvest, 14, 22
health, 31
hive, 9
honey, 9, 33
icing sugar, 25, 34
India, 10

map, 26-27
maple syrup, 29
marzipan recipe, 34-35
medicine, 10
milling, 16
muscovado, 25
nectar, 9
nets, 33
setts, 13
sugar beet, 21-23, 26-27, 29
sugar cane, 10, 13, 19, 26-27, 29
sugar crystals, 10, 16, 25
sugar loafs, 16, 25
sugar lumps, 25
teeth, 31
trade, 19
wasps, 33
white sugar, 25

Pocket Worlds – building up into a child's first encyclopaedia:

The Animal World

Crocodiles and Alligators
All About Pigs
Animals in Winter
Bees, Ants and Termites
Wild Life in Towns
Teeth and Fangs
The Long Life and Gentle Ways of the Elephant
Animals Underground
Big Bears and Little Bears
Wolf!
Cows and Their Cousins
Monkeys and Apes
Big Cats and Little Cats
Animal Colours and Patterns
Prehistoric Animals
Animal Architects
Animals on the Move
Wildlife Alert!

The Natural World

The Air Around Us
The Sunshine Around Us
The Moon and the Stars Around Us
Our Blue Planet
Mountains of the World

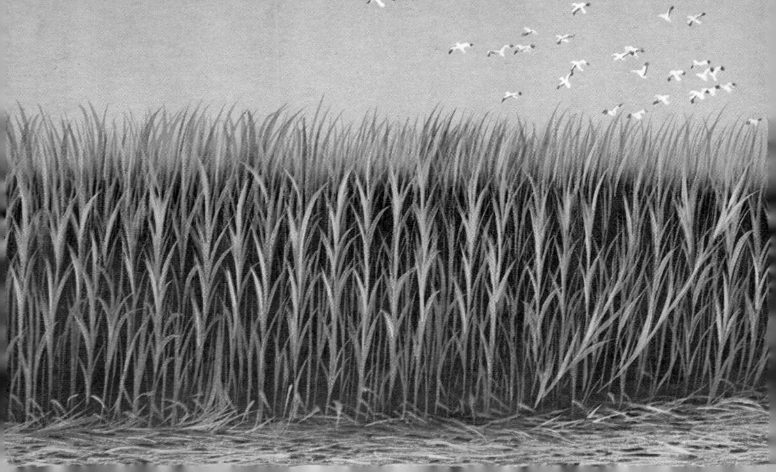